THE CLIVES OF SANDYFORD, STAFFORDSHIRE

CREDO·AMA·ET·REGNA

Alan J. Jones, Canberra 2023

The photograph used on the Front Cover is of the Woodlands, festooned with Spring Jonquils, at the Dorothy Clive Garden (Published, courtesy of the Trustees of the Dorothy Clive Garden).

INTRODUCTION AND ACKNOWLEDGE MENTS

This monograph is a sequel to the book: *13th Corps, A Short History – 1st Battalion North Staffordshire Rifle Volunteers (1858 – 1908)* self-published as a Kindle Book in early 2022 by this author.

The current focus is on a second family - the Clives (the other earlier family being the Williamsons from the period of the foundation of the Corps) that played dominant leadership roles in the 13th Corps, especially from the period 1867 – 1908, but with continuing military and civic engagement through to 1963.

The mere mention of the "Clive family" immediately provokes investigating possible linkages to the well-known "Clives of Shropshire", particularly linking to the Parliamentary representative (MP) for the County of Shropshire (Salop) Richard Clive (1693 – 1771) and his wife Rebecca Gaskill (1694 – 1732), whose most famous (perhaps today, controversial) offspring was Robert (1725 – 1774), best known as "Clive of India", later to become a Major General, a Lord, and First Baron of Plassey (Wales). For our purposes it will suffice to refer to this branch of the Clive family as the "Clives of Shropshire" since there exists some mysterious intrigue about another part of the family that will be referred to as the "Clives of Staffordshire". I add this quote in deference to the book written in

1930 by Percy W. L. Adams on the ancestry of his famous Adams Pottery Manufacturing family, which also refers to "The Clives of Sandyford", the area in Staffordshire where the author grew up to adulthood, and considered as part of the Village of Goldenhill, the ultimate HQ for the 13th Corps.

I am particularly indebted to Helen Burton of Keele University Library, who provided me with an outline (under Copyright law) of the Book by Paul Adams: *John Henry Clive of North Staffordshire and his Descendants*, published in 1947; and advised me to wait for Copyright release to occur on 6 December 2022, the 50th anniversary of the author's death. Special thanks also to Jonathan Goodwin, City of Stoke-on-Trent, who read a substantial section of one of my earlier drafts and attenuated some of my tensions about "hidden information"; and to Ken Perkins of the Apedale Heritage Centre, which features the North Staffordshire areas rich industrial and social history through the varied exhibits in their museum – one of these being the Brick stamped – "Clive Brothers, Sandyford".

I want to offer special thanks to a surviving member of the Clive family of Staffordshire family, John Clive, who introduced me to the term "The Legend of the Clives" and initially drew my attention to the fact that one of his Staffordshire Clive ancestors married a Clive from Shropshire in 1907.

Finally, I want to express my sincere gratitude to Ms Kathryn

Robey, Curator/Director of the Dorothy Clive Garden, Willoughbridge, Market Drayton for her generous support in providing several photographs provided in the text and for the background for the "Front Cover" of the text.

The opening Chapter begins with a short review of Percy W. L. Adams 1930 findings, but then moves on to discuss what a living member of the Clive Family suggested as "The Legend of the Clives", which is also very well covered in Percy Adams book of 1947. My efforts draw attention to the family's Military History and connections with Sandyford/Goldenhill.

I also want to note my use of the Heraldic Arms of Clive of North Staffordshire, designed by John

Henry Clive and on record at the College of Arms (London). It is described as:

Arg. on a fess between three wolves heads erased sa. as many mullets of the feld all within a bordure engrailed erminois.
Crest: a griffin statant arg. holding in the beak a mullet sa.

Alan J. Jones, 2023

TABLE OF CONTENTS

Section	Title	Page
Chapter 1	The Clives of Sandyford	17
Chapter 2	"The Legend of the Clives" and the "Clives of Staffordshire".	29
Chapter 3	The Staffordshire Military Heritage of John Henry Clive (1781 -1853)	43
Chapter 4	Henry Clive (1810 – 1865) – Descendants in Pursuit of Military and Religious Interests	55
Chapter 5	Colonel Robert Clement Clive, V.D. (1846 – 1930).	67

Added note on
John Henry Clive
(1886 – 1932)

Chapter 6   Colonel Harry   78
               Clive (1880 –
               1963), CB, OBE,
               DL, TD.

Postscript   104

Bibliography   105

About the   107
Author

Other Books.   110
Contributions
to Books by
the Author

LIST OF ILLUSTRATIONS

Plate	Title	Page
1	Newfield Hall – The Seat of the Smith Child Family and home of John Henry Clive 1813 -1824. From Percy W. L. Adams Book (1947).	22
2	The Clive Brothers Marl Works, shown on the 1890 Ordnance Survey Map.	26
3	The Clive Brothers Fire-Brick, Courtesy Apedale Heritage Centre	28
4	Birth Registration of a child to Richard and Sarah Clive, 1741, St	34

Bartholomew's,
London

5   Portrait of John Henry   53
        Clive (1741 – 1853)

6   Portrait of Mrs Sarah   54
        Simpson (ex-Clive)

7   Photograph of Lt-   67
        Colonel, Honorary
        Colonel Robert
        Clement Clive (About
        1930)

8   Portrait of Brevit   78
        Colonel Harry Clive,
        CB, OBE, DL, TD in
        retirement

9   Visit of Duke of   90
        Gloucester to
        Staffordshire.
        Standing at the left
        facing the Duke are
        the Lord Lieutenant of
        Staffordshire (Earl of

Harrowby) and
Colonel Harry Clive,
commanding the
Staffordshire Infantry
Brigade

10   Ms Dorothy Clive   92
        (Daughter of Colonel
        Harry and Mrs
        Dorothy Clive)
        Wedding at Maer
        Church, 1938.

11   An Aerial View of the   97
        Colt Bungalows and
        Elds Gorse on the
        Dorothy Clive Garden
        Estate.

12   The Waterfall in the   99
        Dorothy Clive
        Garden.

13   Part of the floral   100
        display at the Dorothy
        Clive Garden.

14   Photograph of   101
     Dorothy Hilda Clive
     and her favourite
     Dog.

CHAPTER 1

THE CLIVES OF SANDYFORD

I first came across the expression "the Clive's of Sandyford" in a book written in 1930 by Percy W. L. Adams on the ancestry of his famous Pottery Manufacturing family, in which he describes a "coterie" of friends in and around Greenfield (the word coterie is not much used these days, so I explain it is "a small group of people with shared interests or tastes, especially one that is exclusive of other people"). The Greenfield Estate was just off what we know today as Furlong Road, down the hill from Christ Church, Tunstall.

The coterie of friends included the Childs of Newfield, the Clives of Sandyford, the Cartliches of Goldenhill, Anthony Keeling; the Henshall Williamsons and the Adams family of Greengates. Percy Adams describes the link between Admiral Smith Child, who on his death in 1813 requested the Clives to take on his grandson while still in his minority and remain as the occupants at Newfield. He also notes that the Cartliches were Master Potters who had acquired a good deal of property around Tunstall; that Keeling was the son-in-law of Enoch Booth, the well-known Potter, (and I add collector of historical information on the Potteries); that John Henry Clive was "not of Staffordshire", but sprang from the same family of Shropshire as that of the first Lord Clive; and that in 1790

Admiral Smith Child had let his "manufactory" to William Adams (born 1746) to work in conjunction with his Greengates works. After the death of Adams in 1805 the Newfield manufactory was again let and operated under the name Child and Clive.

We also learn from the writings of Shaw in 1829, that Newfield included "valuable mines of coal", which were increased in value during the minority of Admiral Smith Child's grandson under the care and management of John Henry Clive. It was the pairing of young Child and Clive that also worked the mines under the same name at Clanway Collieries, a stones-throw from Newfield.

The Newfield Estate was purchased by another William Adams (born 1795) in 1858, and remained in the Adams family at least until 1930, although some portions were sold off, notably the Newfield Pottery was sold by a William Adams (born 1833) some thirty-five years earlier. Most of us living today would remember this factory as Alfred Meakin's Newfield Pottery, and the Meakin's Cricket Ground in Sandyford, which supported a couple of Tennis Courts, a Football Pitch where the Goldenhill Wanderers often appeared, and a youth club equipped with Table Tennis and Snooker Table.

In his book of 1930 Percy Adams also provides a short overview of the effective founding father of the "Clives of Sandyford", and in reality the "Clives of Staffordshire" bearing the name John Henry Clive (born in about 1780).

Adams writes that *"he came with his widowed mother Mrs Sarah Clive to friends at Longton Hall (the Heathcoates). There she married (in 1793) Charles Simpson, Master Potter of Lane End, Longton (son of Daniel Simpson, Master Potter of the same place) at one time in partnership with brothers Turner (sons of John Turner, the eminent Potter), the firm's name from 1803 – 1806 being Turner, Glover and Simpson. The partnership broke up in 1806 and Charles Simpson moved to Goldenhill in about 1810. Charles Simpson is described "as of Newfield in 1824", and his stepson John Henry Clive (then aged 43) as of "Chell House".* During the period of the Child/Clive partnership the whole of the Newfield property was said to have been greatly improved by the judicious management of J. H. Clive, Esq; also *"one of the earliest and most successful introducers of ornamental engraving into the Blue Printing Department of Pottery"*, a trade he learned from Turner.

Plate 1 *Newfield Hall – The Seat of the Child Family and home of John Henry Clive 1813 - 1824. From Percy W L Adams Book of 1947*

John Ward's outstanding *History on the Borough of Stoke-on-Trent*

(1843) describes the appointment, effective from March 21, 1816, of a prominent citizen of Tunstall as the first Chief Constable of that town, John Henry Clive, Esq. Ward notes that John Henry Clive had distinguished himself as a citizen of Tunstall following his responsibility for the planning and erection of a building for public purposes, the laying out of a Market Place, establishing a Market, promoting a public subscription of shares (£25) among the inhabitants and owners of property that entered into an agreement with the then Lord of the Manor, Walter Sneyd, for the purchase of a piece of land called *Stony Croft*, and erecting a Town Hall, or Court House, combined with an adjacent Market area. The Town Hall in question was replaced by

the Clock Tower in 1891/92, and the current Town Hall constructed between 1883 - 85 at the opposite end of the market place to the original building.

During the period described by Ward, a number of the families listed in the opening remarks (Chapter 1) are shown by Ward as also owning and operating coal mines, including H. H. Williamson at Little Pits; Messrs Child and Clive at Clanway; Messrs W. and E. Adams at Greenfield; and Messrs Joseph Heath and Co., at Botany Bay. (I was equally shocked as you may be to read Botany Bay).

I have formed the view that the real origin of the term "Clives of Sandyford" comes from the leasing of land for a brick works by the grandsons (Robert

Clement and William Boulton Clive) of John Henry Clive in 1895. The area of the Clive Brothers, Brick and Marl Works is variously described as at Sandyford, Newfield or Tunstall. The shape of the planned workings is provided in the papers for the lease, and corresponds to an area in a National Survey Map located slightly North of the top end of Cartlich Street.

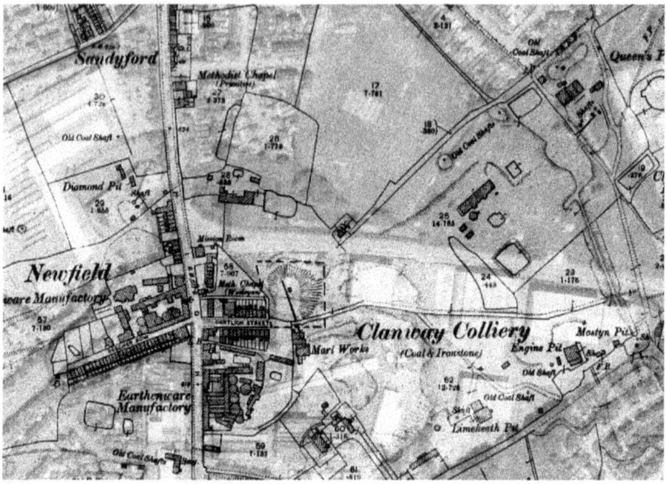

Plate 2 Ordnance Survey Map of the 1890's, National Library of Scotland, showing the Clive Brothers Marl Works and "Marl hole", marked as red area at the end of what was Cartlich Street, Sandyford, Stoke-on-Trent – identified by comparison with the "plan" shown in the Lease Agreement of 1896. The green underlay is of the Map of the present day

To emphasise the "place" of manufacture, the fire-Brick products display the name and place: "Clive Brothers, Sandyford". Robert Clement Clive was the Director of this company and was still running it right until the day he died in a tragic road accident following his departure from the works in 1930. It later became known as the Newfield Brick and Marl Works. In the post-War period, the Clanway Brick and Marl Works developed on the site of the Clanway Farm that we called "Katie Beans". This was ultimately operated by Berry Hill Brickworks of Fenton.

Plate 3 *Photo courtesy of Apedale Heritage Centre, Chesterton, Staffs, photographed by Ken Perkins.*

I add that the demarcation line between Sandyford and Newfield sometimes appears a little blurred, and some 19th Century maps do show the Newfield name running as far south as Furlong Road, Tunstall.

CHAPTER 2

"THE LEGEND OF THE CLIVES" AND "THE CLIVES OF STAFFORDSHIRE

It is clearly accepted that John Henry Clive was the head of a new dynasty of Clives that made their homes in Staffordshire; and there is no doubt that he came to Staffordshire with his widowed mother somewhere between 1781 and 1793 when he was 12 years old. He benefited greatly from his time with his stepfather Charles Simpson and his pottery partners; and more so from his early partnership working in pottery manufacturing with Admiral Smith Child at Newfield, ultimately to become a self-made man in his own right.

The diversity of accolades poured upon John Henry Clive (JHC) are numerous and include his contributions as an engraver for Blue-pottery ware decoration, the building of the first Town Hall in Tunstall, his invention of short-hand, his design and building of a large scale model of a suspension bridge shown at the Great Exhibition of 1851, his invention of the double sextant for measuring lunar distances used in ship navigation, his design and installation of a unique drainage system into the canals for the Clanway coal mines, his writings about the North Staffordshire dialect, and his founding and ownership of the Theatre Royal in Hanley. Added to this, he was also a very successful businessman/property investor.

However, we need to pause here to discuss the relationship of John Henry Clive to the Clives of Shropshire. There have been many other researchers who have tried to probe the genealogical information of the Clives, in particular the work of Percy W. L. Adams in his 1947 genealogical review of John Henry Clive and his descendants, originally commissioned by JHC's Grandson Robert Clement Clive, and expanded later by the next generation, namely, Harry Clive who lived until 1963. A more recent book "*People of the Potteries*" edited by Denis Stuart and published in 1985 simply adopts Percy Adams' (1930) findings:

John Henry Clive was born in Bath on 20 March 1781, the son of Richard Clive, a younger brother of Robert Clive of India. The Clive family shunned his mother Sarah, who took her son to live with the Heathcoates of Longton Hall. Richard, her first husband had died in a debtor's prison, and Sarah married Charles Simpson of the firm Turner, Glover and Simpson in 1793. Her son apprenticed to Turner. The family moved to Newfield where JHC joined Admiral Smith Child in running the Newfield Pottery, later trading from 1811 to about 1828 as Child and Clive.

In my searches through the genealogical records of the Clive family it is evident that the senior Richard Clive (1694 - 1771) was obsessed by a desire to produce a child bearing his own name with his wife Rebecca (1694 – 1745). The family records clearly show multiple births for a "Richard". The first Richard born 1723, died in 1726; a similar 3-year life-term for a Richard born in 1736; and a third Richard born in either 1731 or more likely 1741 appears to relate to a death in Germany in 1763. It is of interest that several of the children born to Richard and Rebecca were recorded at St Bartholomew's in London, the consequence of Richard living in London as the Parliamentary representative for Salop.

Recently, however, I located a birth record for a child to a Richard and Sarah Clive, born on October 13 and baptised on 11 November 1781, recorded at St Bartholomew's in London. However, the child bears the name Richard. This led me to ask the question, could this information be part of the missing link concerning Richard Clive (father) with Sarah?

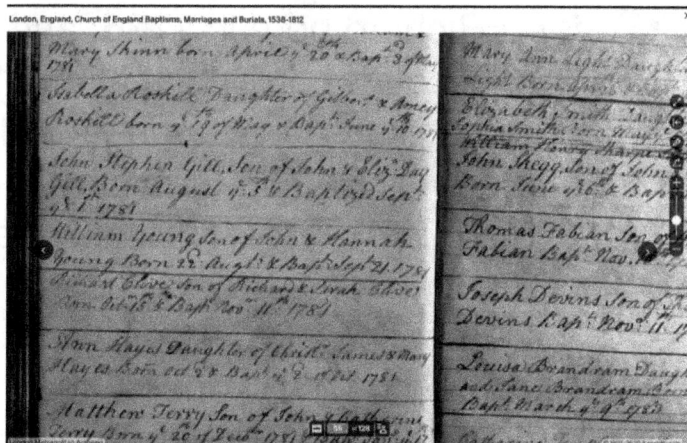

Plate 4 Registration of Birth of a child to Richard and Sarah Clive in 1741, St Barthlomew, London

The Richard Clive involved may have been the son of Richard and Rebecca born in 1741, his life extended beyond the German record of death reported in 1763. Sadly, there appear to be no records of the death of this Richard post 1780 when he fostered the child born in 1781. However, there is a Land Tax record for Richard Clive in Grosvenor Street, London in 1770 only I year before the death of the Richard Clive M.P., so this entry might be for the father of the child born in 1781?

Why do I pose the above questions? Simply, there is scant information on Sarah Clive (widow) until she marries Charles Simpson in Newcastle-under-Lyme in 1783. Sarah was clearly known to the Heathcoates of Longton Hall, hence their providing refuge for her and her son after she was "shunned" by the Clives of Shropshire for one reason or another. In those days it was also quite common-place for an unmarried mother to adopt the name of the child's father, for reasons of respectability, and following separation call themselves a widow.

Perhaps it was such an act that gave rise to the "shunning" of Sarah by the Clives of Shropshire, or was it the shame of apparently having a bankrupt son who went to debtor's prison being all too much for the senior Richard, who after all was the Parliamentary Bankruptcy Commissioner from 1758, and the father of the highly successful Robert, Clive of India.

Many earlier researchers along with myself have attempted to delve more deeply into the principal players involved in this mystery only to find there is surprisingly little information available. First, there is no apparent information on the marriage of a Richard Clive to a

Sarah. In addition, there is little we can find out about Sarah prior to her marriage to Charles Simpson, though that record does tell us that she was 44 years of age when that marriage took place, making her date of Birth about 1749. She lived a comfortable life in Chell until she was 84, largely under the protection of her son John Henry Clive. However, for Sarah no maiden name, or personal birth certificate seem to be available.

Similarly, even following intense searching of the archives of the British Press there appears to be no information to link a Richard Clive over a quite lengthy period around 1781 (the birthdate of the child "Richard" or alternatively "John Henry Clive") to any of the stated misgivings about him – his being in debt, a bankrupt, in debtor's prison, or death in debtor's prison. Finally, no one earlier appears to have located the birth certificate for John Henry Clive, originally suggested as born in the Bath District in 1781.

Perhaps the shame of having a family member designated as a debtor was all too much for the Clives of Shropshire regaled as they were through their son Robert "Clive of India", the first Lord Clive. Percy Adams was the first to suggest some form of cover-up. Certainly, the stories of both Richard and Sarah are shrouded in mystery until Sarah's marriage to Charles Simpson in 1793.

The main elements of conjecture concern the missing official certification on the birth of the mother and the life and death of the father, even the birthplace and date of birth of the child must be brought into question. Though what we do know is that John Henry Clive was to become a very distinguished citizen, and the patriarch of the family, "Clives of Staffordshire".

We can speculate around all manner of scenarios from the "shunning", associations between the "players", the life of Richard Clive (Junior), the life of Sarah Clive up to her marriage to Simpson. Here are a few that have been considered, but to no avail:

Perhaps another male member of the Clive family had a liaison with Sarah prior to her return to Staffordshire. The date of birth of the child (1781) is tell-tale in the respect. It could not have been Robert of India who committed suicide in 1774, or George who died in 1779 (was he Benjamin's son anyway?). Other possibilities rest with Frances (died 1798), William (died 1825) or Nathaniel, whose date of death has not been located.

We cannot be certain that the Richard and Sarah Clive cited in the St Bartholomew Records of 1781 are from the Shropshire Clive family line, though it seems highly probable.

We cannot exclude the possibility that John Henry Clive may have been an illegitimate child?

We can, however, exclude direct family links between Sarah and the Heathcoates, since there was no Sarah Heathcoate born in that generation, and Sarah is not included in the Will of Sir John Edensor Heathcoate.

The St Bartholomew's record of a Richard and Sarah Clive having a child in 1781 seems compelling, despite the differences in date of birth and the suggested name of this child. On this basis and the life and times of John Henry Clive I have come to regard the family link between the Clives of Staffordshire and the Clives of Shropshire is real. Further, even if the early link were proved not to be through Richard, the younger brother Robert Clive of India, the link was later firmly established in 1907 when Harry Clive

(Staffordshire) married Dorothy Clive (Shropshire).

Whatever, "The Legend of the Clives" lives on.

If readers can suggest any evidence to fill the holes in the present analysis please let the author know.

CHAPTER 3

THE STAFFORDSHIRE MILITARY HERITAGE OF JOHN HENRY CLIVE (1781 – 1853)

This Chapter is designed to address research carried over the later months of 2022 on those members of the "Clives of Staffordshire" that featured in the history of the 13th Corps and beyond. It is NOT intended to be a comprehensive review of the entire genealogical tree of John Henry Clive's descendants.

I completed my own independent investigations concerning the critical 13th Corps members, most notably Robert Clement Clive,

Harry and Lawrence Clive in the period from early 2021 to mid-2022. Thus, I was moderately surprised on the day of 6 December 2022 to locate a digitised copy of the book by Percy W. L Adams published in 1947: "John Henry Clive 1781 – 1853 of North Staffordshire and his Descendants". Percy Adam's book had been much sought after by me for some time, but was protected by Copyright up to the 50th anniversary of the death of the author[1]. Adam's book proves to be incredibly comprehensive in dealing with the life and times of John Henry Clive, though it also reflects on the mystic of John Henry's origins implied in my earlier description of the "Legend

---

[1] One month after finding the book on the Web in searching for any records of John Henry Cive as a Volunteer Infantry Officer (1803), the Web listing remarkably disappeared, but I had secured my copy.

of the Clives". Adam's clearly held the view that JHC was born in Somerset (Bath or Bristol), but was unable to substantiate this through any birth registration details. On the other-hand in his later years JHC chose to live in the Bath district, where he and a number of family members are interred. This observation may add credence to JHC's earlier links with that district. In any event it is quite clear that JHC never spoke about his father to any of his descendants, thus maintaining the mystery of his origins".

A short anecdote from Mrs Ann Harvey (daughter of John Henry Clive) is outlined in Percy Adams 1947 book. She commented that her father would seldom talk of his father, but say to family enquiries: *You belong to the best in the land, but it does no one any good to know it. Each of us must make their own way in this world.*

Percy Adams' book on JHC draws on the earlier historical review by John Ward in 1843 - "History of the Borough of Stoke-on-Trent" in describing the Napoleonic threats of 1803. Adams suggests that JHC, at the age of 22 years, was in command of one of the volunteer companies, possibly one of the four created in the Lane End (Longton) district. Adams notes that JHC's daughter Anne (Harvey) provided a reminiscence of her father's role as a volunteer army officer who in recognising that his charges had difficulty in distinguishing right and left provided a solution involving tying hay and straw round the right and left feet, respectively, His commands were thus: "Hay about turn", etc.

In a more recent book (2016) on the 1803 volunteers ("Called to Arms") Paul Anderton notes the difficulties, through shortage of efficient officers in Stoke-on-Trent, and the premature failings in attempts to establish the Tunstall and Burslem Corps. It appears that Anthony Keeling chaired a "Call to Arms" meeting in Tunstall and simultaneously John Davenport proved to be the mover and shaker for the formation of four companies, each of 80 rank and file, in Lane End (Longton), which absorbed some of the recruits from Tunstall and Burslem. Other officers discussed by Anderton are Samuel Cartlitch, Charles Simpson, E. and G. Goodwin, T. Cartlitch and George Reade; and, in particular, he notes John Glibert, who successfully established a Volunteer Corps at Clough Hall in Kidsgrove.

I have attempted to comb all the sources of Army Lists of that time, notably held by the British Library, but have been unable to locate any reference that John Henry Clive was in command of a Corps of infantry in Staffordshire. Unfortunately, the existing lists focus on the Commandants of combined district companies (corps). The British Library List of 1804 comprises lists of the Volunteer and Yeomanry Corps of the United Kingdom. To which is added an appendix containing the complete Regulations for the Volunteer Establishment with an abstract of the Consolidated Volunteer Bill. London: J. Stockdale, 1804. Library reference **8824.b.26.** A second List of the Officers of the Gentlemen and the Yeomanry cavalry, and Volunteer Infantry of the United Kingdom. pp.830 [London], 1804. 8° **BS.45/150** is also held at the National Library of Australia. Again, only the names of the Commandants are listed, that is: Newcastle – J. H. Northern; Clough Hall – John Gilbert; Hanley and Shelton – James Whitehead; Stoke, Penkhull and Fenton – D Walley; Tunstall – Samuel Cartlitch; Longport – John Davenport; Etruria – John Wedgwood; and Lane End – William Turner.

The excellent historical record: *Years of Victory 1802 – 1812*, by Arthur Bryant published in 1944, barely mentions the volunteer movement created to defend the homeland, although the concluding Chapter: *Over the Hills and Far Away* does speak of "England becoming too strong for Napoleon". Even so Napoleon did muster forces of 80,000 men at Boulogne in 1811, though with little chance of crossing the Channel.

Further, the Percy Adams book of 1947 does draw on numerous papers and notes written by John Henry Clive himself, not least of which point to John Henry Clive's own interest in the genealogy of the ancient Clive family, and including detailed analyses of several of his notable published inventions. I particularly reiterate the subjects of the suspension bridge design exhibited at the Great Exhibition in 1851; his analysis of the North Staffordshire dialect; his navigational sextant; and his development of

Shorthand, though his approach was later superseded by that of Pitman. Another area that appears to have been taken up by other writers relates to the "drainage scheme" for coal mines; notably at the Clanway Collieries, which belonged to the Childs, and in part the Sneyds (of Keele), and were worked by Child and Clive, and later by the Clive family alone. Although commented on earlier in this text as a positive outcome, it seems that John Henry Clive's Diary of 1853 suggests that the scheme never materialised.

On the personal side of his life John Henry Clive married twice, first to Lydia Cash in 1809, and secondly to Elizabeth Billington (ex Roylance) in 1824. His first marriage fostered five children, the youngest Robert Clement died as an infant aged four years, and his eldest son Charles William (1807 – 1839) remained unmarried and died at the early age of 32 years, John Henry and Lydia's second son Henry (1810 – 1869) was to succeed as head of the family and promulgate the "Clives of Staffordshire" line in the generation that followed. John Henry and Lydia also had two daughters Mary (Meir) (1815 – 1899) and Ann (Harvey) (1815 – 1899). JHC's marriage to Elizabeth Billington (ex Roylance) provided two daughters, Elizabeth Roylance (1825 – 1894) and Lucy (1828 – 1907). It is clear from Percy Adams' Book that Ann Harvey was also a good source of anecdotal evidence on the history of the Staffordshire Clive family.

Plate 5 *A Portrait of John Henry Clive*

Plate 6   *A Portrait of Sarah Simpson (ex Clive)*

CHAPTER 4

HENRY CLIVE (1810 –1865). DESCENDANTS IN PURSUIT OF MILITARY AND RELIGIOUS INTERESTS

As mentioned earlier, Henry Clive (1819 -1865) was the second son of the union between John Henry Clive and his first wife Lydia Cash. It became Henry's responsibility to take some of the business opportunities established in Coal Mining between John Henry Clive and the Childs at Clanway. He also established Henry Clive and Co., of Clay Hills and was a dealer in lime mined at the Clanway site as early as 1841, and was also a brick and tile manufacturer around 1852. Henry married Anne Hancock of Liverpool in August 1842. They lived in Broomhill, Tunstall. Percy Adams notes that Anne's father was a Captain in the Merchant Navy who carried "Letters of Marque", a commission by the British Government to make reprisals on vessels of other countries, notably Russian, Swedish, and Danish ships that had formed "the League of Armed Neutrals". When I think of it, Britain had been good at Piracy at Sea for several centuries.

As well as being in mining and manufacturing, Henry was a member of the Board of Health of Tunstall. According to Percy Adams, Henry's father had presented him with a very nice little property at Broomhill, which commanded a good view of Chatterley and Bradwell Wood, but 4 years after the death of his father he moved to Moor House in Biddulph, where he died in 1865. His wife stayed at Moor House only for a few months and over the next 20 years she moved about 6 times before ending up in Clanway Cottage in 1885, where she died in 1889. The couple are buried together in the Churchyard at Biddulph.

Henry and Anne had 8 children, the first named after his Grandfather John Henry Clive (1843 - 1871), who became a mining engineer. Their second son George Roylance Clive (1844 – 1875) migrated to Australia and married Ellen Diston in January 1875. He died there in the following November; Robert Clement Clive to whom we dedicate Chapter 5, as a key leader in the 13th Corps and the entire 1st Battalion; William Bolton Clive (1847 – 1920), who became a partner in business with his brother Robert Clement; Herbert Clive (1859 - unknown) who apparently died in Australia; Stephen Clive (1855 – 1927), who became a Master Potter in Tunstall to about 1880 when he moved to New Jersey USA, and married Helen Jackson dying in New Jersey in 1889. However, we have a record of his unassisted passage for Stephen to Australia in 1878 at the age of 23 years. Finally, Francis Biddulph Clive (1857 – 1919), educated at St John's College, Cambridge. He became a Cleric in holy orders, a Deacon (1884), a Priest (1885), and a Curate in Bodalla, New South Wales, Australia, and later in North

Queensland. He died in Rockdale, New South Wales in 1919 at the age of 62 years, and was buried at Lake Bathurst, Goulburn, New South Wales.

I was pleased to find an article in the Staffordshire Sentinel of February 12, 1875 which demonstrates some of the unity of interests in Henry Clive's family. The article describes a "Soire at the National Schools Meeting" held in the Christchurch school premises, Tunstall, where an audience of 60 – 70 finely attired ladies and gentlemen were entertained by singers Messrs R. and H. Clive accompanied by pianist MIss Turner. At 9.30 pm the Vicar addressed the audience commenting on the elaborate decoration of the stage, which featured a portrait of Shakespeare at the centre, and at the sides, on pedestals, busts of Moliers and Racine; and he also commented on the excellence of the entertainment offered by the Clives. A light supper was served, followed with more entertainment with Mr Stephen Clive singing a version of "Old King Cole". and Mr Shorrocks the Burn's songs "Afton Waters" and "Of all the Airte". The evening concluded with all singing the National Anthem.

I am also intrigued by the fact that four members of Henry's family found their way to Australia. It is noteworthy that George Roylance Clive married a young female from the area of Fiery Creek, which was part of the Victorian Gold Rush. Gold was discovered near Beaufort in 1852 in the tributaries of Fiery Creek, and north of Beaufort in Fiery Creek from 1854. The population on the fields was 50,000 in 1855 and reportedly reached approximately 100,000 people at its height in the late 1850s and produced 450,000 ounces of gold over a two-year period, 1855–1856. The Fiery Creek gold rush dissipated by 1859, though Gold-dredging went on to about 1918.

It is possible that Herbert Clive may have also have found his way to another gold rush area, that of Winneshiek County in Iowa in 1870, where he is noted "in residence", at least for that year. However, he migrated to Australia arriving in October 1875. Unfortunately, the only sources of Herbert's activities in Australia are reports in the Victorian Police Gazettes of 1877:

Herbert Clive is charged, on warrant issued by the Tarnagulla Bench, with obtaining money and goods from Robert Anderson and Co., Tarnagulla, on the 24th instant, by means of a valueless cheque for £4. Description: English, about 26 years of age, 5 feet 9 inches high, fair complexion, small fair moustache only, blue eyes, round shoulders, snuffles when speaking; dressed in grey tweed trousers and vest, blue striped cotton shirt, and black felt hat, and carried a black satchel with strap over his shoulder. He represented himself as travelling for his brother, who, he said, owns a pottery in Staffordshire. He was seen in Sandhurst on the 13th instant. He is known *at the Princess of Wales and Corio Hotels, Ryrie Street, Geelong. This offender obtained money from William Heffernan and William Atkinson, hotelkeepers at Sandhurst, on the 12th and 13th instant, by means of valueless cheques, He went to Echuca by the 8.20 pm train on the 13th, Cases: O.1704 and O.1770, 20th March 1877.*

Herbert Clive obtained £2 10s from Harry Franklan, baker, Broadmeadows, on 20th ultimo, by means of a valueless cheque. No warrant. This offender arrived in Echuca by first train from Deliliquin on the 20th instant, and obtained £4 from James Speary by means of a valueless cheque. He left for Melbourne by train, saying he was going to Sydney. He was dressed in short silk coat, drab felt yankee hat, and light tweed trousers. On the 27th instant he uttered a valueless cheque for £2 17s 6d to Louis Levy, Flagstaff Hotel, Dudley Street, West Melbourne. Cases O.1829, O1944, O1946. 27th March 1877.

Herbert Clive, charged with obtaining money by means of valueless cheques from Robert Anderson and Co, William Heffernan and William Atkinson, Harry Franklan, James Speary, Louis Levy and Eliza Palmer has been arrested by Melbourne detective police. 10th April 1877.

I have no follow-up on Herbert Clive from that point, and can find no record of his death. I suggest that Stephen Clive's visit to Australia in 1878 was probably to catch up with his older brother, even to bail him out of goal. At about that time, Captain Stephen Clive was in command of the Tunstall Rifle Volunteer Corps (SS., 26th May 1877).

Now more attention on the Military linkages of the more familiar descendants of the Clives of Staffordshire most being associated, initially at least, with the 13th Corps (North Staffordshire Rifle Volunteer Regiment).

CHAPTER 5

COLONEL ROBERT CLEMENT CLIVE, V.D. (1846 -1930).

Plate 7 Robert Clement Clive
Lt. Colonel, Honorary Colonel
(From the Evening Sentinel Newspaper of 15
April, 1930)

In my earlier monograph: *13th Corps, A Short History – 1st Battalion North Staffordshire Rifle Volunteers (1858 – 1908)* I described an officer as Robert Clive (1846 – 1932). His full name is Robert Clement Clive, born in Tunstall in January 1846. He joined the Kidsgrove and District Corps as an Ensign in 1867, became Captain of the 13th Corps in 1876, and was promoted to Major in 1888 before moving to the Tunstall Corps. He appears to have stayed as Captain with the Kldsgrove Corps during its most trying economic crisis, the years of the long depression, that saw many locals out of work, bankrupted and destroyed mining businesses, and culminated in the move of the 13th Corps headquarters to Goldenhill.

Robert Clement Clive rose quite rapidly through the Officer ranks to become the Commandant Lieutenant-Colonel of the entire 1st Battalion of the Regiment (of 1500 men) by 1900, having been second in command from 1892. He was granted the title of Honorary Colonel in 1890. Following the end of the Rifle Volunteer movement in 1908 he became the Commanding Officer of the 5th Battalion of the North Staffordshire Regiment, and at the time of his death he was Commander of the Staffordshire Infantry Brigade – a military career spanning over 63 years.

The main reason for expanding the story on Colonel Robert C Clive arises from a front-page story in the Staffordshire Sentinel of Tuesday 15 April 1930, which describes the tragic death of the

Colonel "as the result of injuries received in an accident at Goldenhill". The report goes on to say that at about 4.30 pm on the previous afternoon Colonel Clive left his Marl Works at Tunstall, his intention being to board a motor-bus. He was knocked down by the vehicle resulting in severe head injuries. After receiving medical care, he was taken to hospital (North Staffordshire Royal Infirmary), but lived only three hours after admission.

Let me describe more about the man, not just his long service in the Military. Colonel Clive was first an astute businessman, a servant to local government including being a Chief Bailiff, a Magistrate, President of the North Staffordshire Royal Infirmary, and a devout practising Christian. He was church warden at Tunstall,

Lawton and Hartshill, and trained choirs at all those churches, and at Rode Heath, He made a special study of church music and was an accomplished musician. Most of all, he was regarded as a "typical English gentleman" by all that came into contact with him.

On the business side, Colonel Clive initially managed the Clanway Colliery, was head of the of the firm of Clive Brothers, Newfield Brick and Marl Works, Tunstall; was a Director of J Gimson and Co (1919), Fenton, and Chairman of Directors of Furlong Mills Co Ltd., Burslem. It was his role in the Marl works with his brother William Boulton Clive that gave rise to the term Clive Brothers, Sandyford, as noted in Chapter 1. I was somewhat amused by the prices of bricks outlined in the following advertisement of August 1878 provided by the Apedale Heritage Centre: *For Sale, good Hand-made Red Bricks from 19 shillings to 20 shillings per 1,000. – Apply, Clive, Son, and Myott, Clanway Colliery, Tunstall.*

From about 1871, the Colonel began his connection with local government. For example, he was elected a member of the Tunstall Local Board and appointed Chairman and Chief Bailiff in 1873, serving on what became the Tunstall Council until 1902. In 1873, he also was elected as a member of the Wolstanton and Burslem Board of Guardians serving in various capacities until 1890. Prior to his work on the County Council he was a member of the County Police Committee and County Asylums Committee.

In 1880, the Colonel was made a Justice of the Peace and succeeded to be Chair of the Pirehill North Division in 1894. In all his magisterial work it is said that "he revealed a fine judicial mind, and was in full sympathy with the modern spirit in the administration of justice". It is said that he also displayed a deep interest in the welfare of the police and was held in high regard in the Newcastle Court, He had presided over that Court on the very day of his death.

Here I revert back to the first-born brother of Henry Clive (Chapter 4) – in fact the second Clive of Staffordshire bearing the name John Henry Clive (1843 – 1871). He married Charlotte Hannah Wedgwood in 1865 and their son was the third in descent to bear the name John Henry Clive (1886 – 1932). He attended High School in Newcastle-under-Lyme and wanted to join the regular Army, but was discouraged by his mother. However, in 1886 he became a Lieutenant in the 1st Battalion North Staffordshire Rifle Volunteers Regiment and Captain in command of the Tunstall Corps in 1891. He retired from the Corps in 1901 following 15 years of service. He married Amy Cooke in 1910, and moved to live in Orban, Scotland, where he died in 1932. The couple had no children.

CHAPTER 6

COLONEL HARRY CLIVE (1880 – 1963) CB, OBE, DL, TD.

When the 13th Corps was re-established in its new Drill Hall and Armoury in Goldenhill, Captain W. H. "Wally" Grindley was in command from 1896 to 1900. The Military reference books of that time suggested that Captain G. W. Laybourn was Grindley's successor but just recently I discovered the disembarkation papers on Laybourn that show him and Lieutenants H. J. Johnson and E. F. Hodson as three of the officers in a 60-man contingent from the 1st Battalion North Staffordshire Rifle Volunteers Regiment that shipped out on the "Nineveh" in March 1900 to fight with the regular Army in the Transvaal Campaign, Boer War. These men were sent off following a Mayoral Reception put on by all the six-towns of the Potteries (*Staffordshire Sentinel, 20 January 1900*).

Captain Laybourn returned to command the 13th Corps, Goldenhill, in 1902. At some point in early 1904, Captain Laybourn was joined by two assistant officers, Lieutenants Harry and Lawrence Clive, who coincidentally were cousins, Lawrence being the son of William Bolton Clive and his American wife Kate. (I draw particular attention to the spelling of Lawrence since it was his mother's maiden name, though almost all the newspaper reports of the day insisted on using an incorrect spelling "Laurence" for his name). Harry and Lawrence were promoted to the rank of Captain in 1904 by the then Regimental Commander Colonel Dobson, and replaced the "retiring" Captains Laybourn and Alfred Meakin of the Goldenhill and Tunstall Corps, respectively.

It is apparent from some of the "pep-talks" given by Captain Harry Clive to the men of his Corps that he was made of "stern military stuff" (I suspect under orders to demand a more professional approach to military training to match that of the regular Army). An agenda that would turn the Rifle Volunteers into a much more professional military organisation to serve as Reserves for the regular Army. By 1908 the 13th Corps had merged

Plate 8 Brevit Colonel Harry Clive, CB, OBE, DL, TD.
This Portrait is currently held at the Dorothy Clive Garden
(Published, courtesy of the Trustees of the Dorothy Clive Garden)

with the Tunstall Corps and Captain Harry Clive took that

command. The Staffordshire Sentinel of 25 January 1908 reports on a "highly fashionable" event held at the Town Hall in Market Drayton and included in a range of distinguished guests were *Colonel Robert Clement Clive, Mr and Mrs Harry Clive, Mr Robert Clive, a Mrs Harry Clive and Miss Hilda Clive (Harry Clive's sister)*. Captain Harry Clive had married Dorothy Hilda Clive (of the Shropshire Clives) in 1907 in Kensington, London. Thus, reuniting the two "Clive Lines".

An article in the Staffordshire Sentinel of 13 June 1910 reports that Lawrence Clive (later Captain 1St Battalion North Staffordshire Volunteer Regiment) has been gazetted as a Lieutenant in the newly formed 5th North Staffordshire Regiment. However, there was very little in the Newspaper archives concerning the relationship between the Clive family members and the Military other than the formation of the 5th Battalion of the North Staffordshire Regiment, which included a small infantry group based in Hanley; and a Royal Artillery 1st Staffordshire Battery with Company by town, for example, Company D -Tunstall.

World War I (WWI) was looming and both Harry and Lawrence were called to battle in France. In a badly folded copy of the Staffordshire Sentinel of 19 October 1915 (where some words are buried under the fold) there is a report that Lieutenant Harold Clive, the brother of Lawrence Clive has been wounded in France, sustaining serious injuries. The article also notes that Lawrence Clive had been seriously wounded sometime earlier, For Harold and Lawrence the good news was that Harold's wedding to Annie Vivian Audley at Longsdon Parish Church is reported (SS 12 February 1916), with Lawrence listed as best man. We can attest from the War records of Major Harry Clive that he was also wounded in France on 2 February 1915. All three Clive men were repatriated to the UK.

This coincidence of members of the same battalion being wounded suggests this was most probably in the renowned Hohenzollern Redoubt region in North Eastern France just near the Belgian border. The Germans were particularly well prepared for these early battles. The British forces comprised the 137th, 138th and 139th Brigades of men from Staffordshire; Lincoln and Leicester; and Nottinghamshire and Derbyshire, respectively; plus four Brigades of the Royal Field Artillery; 3 companies of Royal Engineers and the 1st Battalion of the Monmouthshire Regiment – all part of the 45th North Midland Division, The 5th and 6th Battalions of the North Staffordshire Regiment provided approximately half of the manpower of the 137th Brigade of an estimated 6000 men. In a series of phased battles lasting 5 – 6 days at about 3 weekly intervals from August through to the beginning of October the 46th North Midland Division suffered casualties of 3763 men. German casualties amounted to about 9,000 men, but other Divisions of the British army suffered another 2115 casualties. An incredible blood bath that the three Clives survived.

At the conclusion of WWI the King's Birthday Honours List of 1919 included Major Harry Clive, awarded the Order of the British Empire (OBE) with the citation: "In recognition of distinguished services rendered during the War". Things were really looking up for Harry Clive and by December 1919 (SS, 18 December 1919) the War Office decided to reconstitute the Territorial Force of Great Britain, recruited "voluntarily", and utilised only in a state of emergency. Major Harry Clive was appointed temporary commanding officer of the 5th North Staffordshire - "this distinguished local Territorial Battalion". He was the first relatively low-ranked Territorial officer to have such an honour conferred upon him. I contrast his appointment with that of Colonel John Keat who was nominated to take command of the local Field Artillery, and Sir Percival Haywood, DSO, command of the Yeomanry.

Harry Clive was promoted to Lieutenant-Colonel of the 5th North Staffordshire Regiment in 1923, and was in command of the 5th as the Guard of Honour for the Prince of Wales' visit to Stoke-on-Trent on 6 June 1924. On Wednesday 14 March 1928 Colonel Harry Clive addressed a huge gathering at the Tunstall War Memorial, which he unveiled. He uttered the stirring words: "You in Tunstall have always been proud of the part you played in your civic life. You need to be no less proud of the part your sons have played in the making of the Empire's history".

To elaborate on the story of Lawrence Clive's life. His military record at the end of WWI simply records him as a Captain in the Territorial Force Battalion of the North Staffordshire Regiment. In civilian life he was a well-respected inspector of mines, while many of the other Clive relatives held substantial holdings in the coal mines. In the 1911 Census, Lawrence was boarding in a house in Newcastle-upon-Tyne, and his profession listed as a Civil Servant - **M**usic Inspector. (I suggest the **M**-word is for Mining). Following completion of his war service Lawrence's Pension Card records his emigration to Alberta, Canada, having survived "Charcot-Codyre". I believe that Charcot-

Codyre relates to a rare foot condition involving nerve damage sustained in the War. He died in Canada aged 81 in the small seaside town of Campbell River on the East Coast of Vancouver Island, British Columbia. I suggest that Lawrence is the exemplar of a man escaping from the "militarism" of his life. Regrettably, my Ancestry searches exclude Canada and the USA - but something tells me he was happy. I note that his brother Harold Clive (1889 – 1944) served as a Lieutenant with the 5th North Staffordshire Regiment in WWI, was wounded and transferred to Garrison duty in Bermuda. He returned to the UK and married in 1916.

I was again reminded of the Hohenzollern Redoubt of 1915 on reading an article in the Staffordshire Sentinel of 2 August

1926 covering the presentation of the Colours and the War Memorial gifts to the 5th North Staffordshire Territorial Battalion at the Stoke City football ground in the presence of the full Battalion, and a mighty gathering of friends and family, The Editorial notes that the Battalion now possessed the King's Colour and the Regimental Colour, a bugle, a drum, and fife (like a flute) Band, the latter presented in memory of officers that fell in the War. The fife Band was added to the existing Brass Band of the North Staffords. Lieut-Col Harry Clive invited other ladies of North Staffordshire to present further bugles, drums and fifes to complete the Band as a memory to those lost in the War,

The August 1926 article also notes that the Clive family (Staffordshire), including the two

Colonels Robert and Harry, are "kin to that of Robert Clive, the creator of our Indian Empire, who belonged to a very old Shropshire family. In addition, some details of what Lieut-Col Harry Clive did over the period 1916 – 1923 when he was invalided home are included. He was in fact appointed Staff Captain, North Midland Reserve Brigade in May 1916. He was appointed A.1.Q.M.D.5 at the Headquarters of the Northern Command in June 1917; and D.A.O.M.G at HQ Northern Command in April 1918. (I apologise for my inability to translate the Military initials other than Headquarters, as taken from the Newspaper article).

The planning of an ex-Servicemen's Reunion Dinner for men who served with the 5th North Staffordshire Territorials at the Grand Hotel in Hanley is reported in an article in the Staffordshire Sentinel of 29 December 1926. Over 200 tickets had been sold in advance and guests were reminded that Assembly was at 6.30 pm, Dinner punctually at 7.00 pm. The Band of the Battalion was to play during dinner, and a social evening to follow. All old comrades were permitted to wear ordinary civilian clothes, and all the familiar Senior Officers were guests. It seems that the evening was a great success – incidentally tickets were 3 shillings each, which included dinner – *inflation struck thereafter.*

The article that follows is a photograph of an important occasion in 1929, and speaks for itself. Colonel Harry Clive is on the far right of the group.

Duke and V.C.—During his visit to Burton the Duke of Gloucester chatted with Lce./Cpl. W. H. Coltman, V.C. who has won more military decorations than any other N.C.O. Standing behind the Duke are the Earl of Harrowby, Lord Lieutenant of Staffordshire, and Colonel Harry Clive, Commanding the Staffordshire Infantry Brigade.

Plate 9

In 1930 Harry was made Commander of the Staffordshire Infantry Brigade after being the Commanding Officer of the 5th North Staffords for so long. Newspaper articles of 18 August 1931 report Colonel Harry's announcement that he was giving up the command of the Staffordshire Infantry Brigade. The report notes that 30 years had passed since his father had commanded the 1st Battalion North Staffordshire Regiment and that he had fully maintained the soldierly family tradition by his own record while he was in command of the "Fifth" for some 28 years in both peace and war, The article draws an analogy with great soldiers who had held the command before him, notably Brigadier-General J .V. Campbell of the Coldstream Guards, the famous "Tally Ho" VC, who was with the Brigade at the Hohenzollern Redoubt and the crossing of the St Quentin Canal. Colonel Harry Clive told the Brigade in his speech at the break up of a camp in Buxton *"It is nearly 130 years since the first citizen army was formed to repel the threat of a foreign invader. When I leave this Brigade it will be the first time since its formation that there has been no Clive in the citizens army of Staffordshire"*.

Significant among Colonel Harry Clive's achievements were the many functions that he continued to attend well into the 1930's, but none more significant than the Wedding of his daughter Hilda Clive at Maer Church (SS 11 May 1938). I could not avoid providing a Bridal picture since the fashion was almost identical to that worn by my mother at her wedding in the same year

MISS DOROTHY CLIVE'S WEDDING AT MAER :: WOODLAND FIRE HA\

Plate 10

The civilian life of Harry Clive followed very much in his father's footsteps, partly through inheritance. He was appointed a

Justice of the Peace and went on to serve as a Magistrate in the same Court (Division of Pirehill) as his father from about 1921. I took some pleasure in reading about his work as a Newcastle County Licensing Justice. A report (SS, 15 November 1937) addresses one of my old haunts, the Sneyd Arms pub in Keele being granted alehouse approval, and the Rookery Social Service Club in Kidsgrove granted a licence for music and dancing. A short note in a memoir on his father talks of the immense gratification gained by Colonel Robert Clive in seeing that his son Colonel Harry Clive, OBE, TD (Territorial Decoration), had such a distinguished career in the Territorial movement.

In the world of business Harry Clive, like his father was involved in the coal industry, being the joint Managing Director of Parkhouse Colliery in Chesterton, along with his brother Robert, and the Earl of Harrowby. This is reported in the SS 14 July 1947, just before the advent of the National Coal Board. The Colliery Manager was particularly in praise of Colonel Clive's kindness to the staff over the long years they had worked together. He was also Director of the firm Messrs J. J. Gumsom Co (1919) Ltd, which in 1931 amalgamated with Messrs T. Arrowsmith and Sons Ltd of Burslem to become a large supplier of "stilt" and "spur" to the pottery industry (stilt and spur are the spikey ceramic pieces used to separate clay vessels when they are stacked inside saggars for firing).

In his later years Colonel Clive was a keen gardener, showing off his small wild woodland garden on several occasions to enthusiasts travelling around Britain to look at garden exhibits. In 1939 he started to create the Dorothy Clive Garden to delay the progressive effects of Parkinson's disease on his beloved wife. She died in 1942, but after her death the Colonel continued to develop the garden and opened it to members of the public. In 1958 he established the Willoughbridge Garden Trust (an independent charity) to safeguard the Estate. The current description of the Dorothy Clive Garden notes that it is a charitable garden trust, set up by the Colonel as a place of rest and continued horticultural education for the general public inviting visitors to explore and discover the intimacy and

tranquility of the garden.
(https://dorothyclivegarden.co.uk).

By way of background, the marriage of Harry and Dorothy Clive in 1907 saw them move to a property in Willoughbridge, near Market Drayton, Shropshire, about twelve miles from Stoke-on-Trent. They initially lived in an area called Elds Gorse on the property though later the Colonel occupied the Colt Bungalows where he lived until he died. **Plate 11** is an Aerial view of these locations on the property.

Plate 11: An Aerial View of the Colt Bungalows where Colonel Clive lived until he died and
Elds Gorse where he and Dorothy lived when they were married.
(Published, courtesy of the Trustees of the Dorothy Clive Garden).

See Page 97

Some of the magnificence of the Dorothy Clive Garden in bloom is show in the following Plates 12 and 13, and on the Cover of this book.

Descriptors:

Plate 12: The Waterfall in the Dorothy Clive Garden, (Published, courtesy of the Trustees of the Dorothy Clive Garden). See p99.

Plate 13: Part of the Floral Display (Published, courtesy of the Trustees of the Dorothy Clive Garden) See p100.

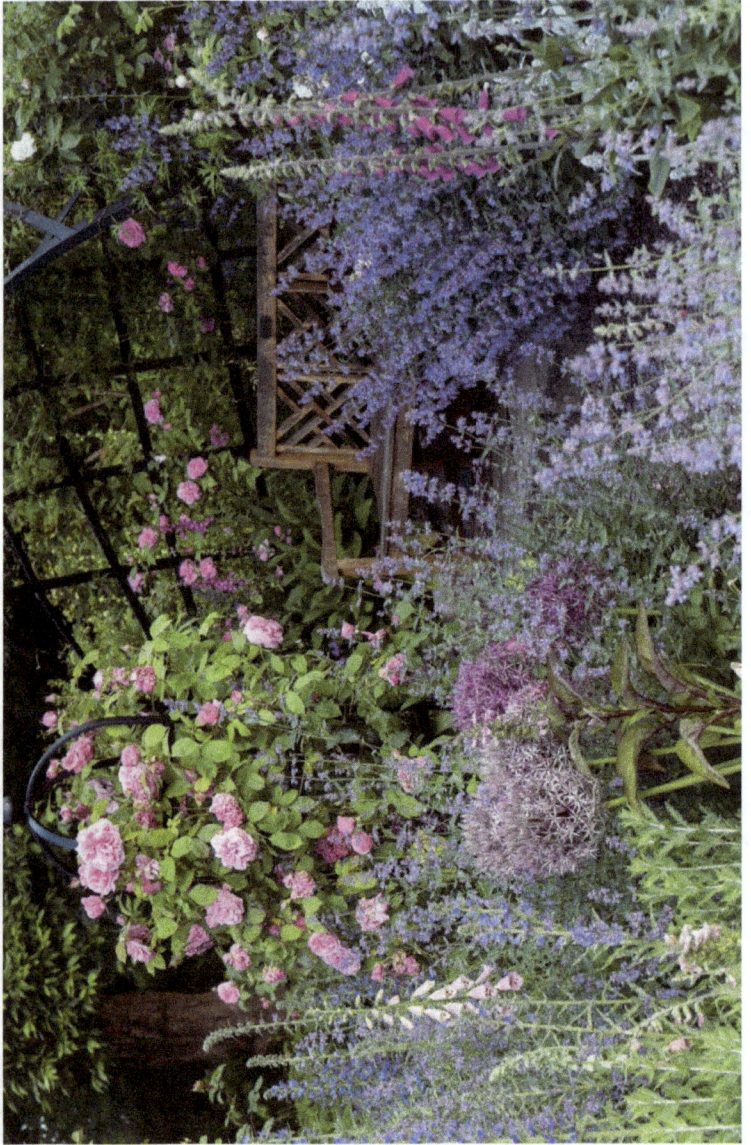

Today, the Dorothy Clive Garden continues to attract many visitors, famed as one of the best rhododendron and azalea showcases in the UK. It seems befitting to show a portrait of the lady to whom this garden is dedicated (Plate 14).

Plate 14 A photograph of Dorothy Hilda Clive and her favourite Dog. (Published, courtesy of the Trustees of the Dorothy Clive Garden).

A further significant event for Colonel Harry Clive is reported in the Staffordshire Sentinels of 9 and 10 November, 1948. It describes three stained glass windows commemorating the 5th Battalion North Staffordshire Regiment that were installed on the main staircase leading to the Jubilee Hall in the Town Hall, Stoke-on-Trent. The handing over of the windows to the City was made by Colonel Harry Clive, CB,, Honorary Colonel of the 576 Searchlight (5th North Staffordshire) Battalion, and accepted by the Lord Mayor. *(It would be valuable if someone living in Stoke-on-Trent would care to view and photograph these windows since a clause in the handover suggested that should Stoke obtain a "new" Town Hall the windows should be moved there – the author would be most interested in seeing a copy of the photographs).*

I draw attention to the title of Colonel Clive, CB - Somewhere near the time of his "retirement" in 1932, Colonel Clive was granted the title of Brevet Colonel (CB) giving him the non-paid rights of a full Colonel. Colonel Harry Clive died on 2 January 1963 at the Stamer House Nursing Home, Hartshill, at the age of 83. It is believed that the Colonel's ashes were scattered to join his wife's ashes in the York Stones area of the Dorothy Clive Garden – "The Garden of Memory" (located near the southern end of the Quarry Garden).

POSTSCRIPT

There may be several other members of the Clive family that had associations with the Volunteer Army. For example, we have located the resignation from the 9th Staffordshire Rifles in March 1876 Captain Arthur Clive Meir; and Lieutenant Stephen Clive from the 13th Staffordshire Rifles (SS., 22 March 1876). I cannot be certain if this is the same Stephen described earlier as a Rifle Volunteer officer. I also note a Robert Michael Clive (1915 –), whose military career was cut short by wounds sustained on D-Day There may be many other Clive family members who participated in Volunteer Military organisations that are not mentioned. If the reader is aware of any missing information of this nature, please let the author know.

BIBLIOGRAPHY

Percy W. L. Adams, *Notes on some North Staffordshire families, including those of Adams, Astbury, Breeze, Chalinor, Heath, Warburton, &c.*, E.H. Eardley, Tunstall, England, 1930. (https://archive.org/details/notesonsomenorth00adam/page/10/mode/2up).

Simeon Shaw, *History of the Staffordshire Potteries*, 1829. (http://www.thepotteries.org/shaw/index.htm).

John Ward, *The Borough of Stoke-on-Trent. Printed and published by W. Lewis and Son, Finch Lane, London, 1843. (Available as a Google book).*

Paul Anderton, *Called to Arms 1803 - 1812 in the Staffordshire-Cheshire Border Region, Audley and District Family History Society, Stoke-on-Trent,* 2016.

Percy W. L Adams, *John Henry Clive 1781 – 1853 of North Staffordshire and his Descendants*", published by G. T Bagguley, Newcastle-under-Lyme, Staffordshire, 1947.

People of the Potteries, edited by Denis Stuart, Published by Department of Adult Education, University of Keele, 1985.

Arthur Bryant, *Years of Victory 1802 – 1812*, published by Collins Clear-Type Press, London and Glasgow, 1944.

ABOUT THE AUTHOR

Alan J Jones, BA(Hons), MSc, PhD, FRACI C Chem has held a number of Teaching and Research positions in universities throughout the world including the University of Keele, UK; Monash University, Australia; University of Utah, USA; Colorado State University, USA; University of Alberta, Canada; the Australian National University (ANU), Australia. He also served as a Senior Diplomat to the Federal Republic of Germany as Counsellor Industry, Science and Technology, Australian Embassy, Bonn from 1990 to 1995.

Alan held honorary appointments as Adjunct Reader, Technology and Innovation Management Centre, University of Queensland (2001 – 2003), and in a similar role at the Business School, University of Auckland, New Zealand to 2006.

Following his official "retirement" he was appointed a Visiting Fellow, National Graduate School of Management, Australian National University, working in the ARC-funded *Australian Systems of Innovation Study (2002 – 2008)*; and Adjunct Professor of Innovation, Management and Performance Management, in the Division of Business, Law and Information Sciences, University of Canberra. He also served a member of the Management Committee of *Chemistry in Australia*, Royal Australian Chemical Institute (2003 – 2014).

Alan's research speciality was in multi-nuclear magnetic resonance (nmr) spectroscopy, particularly carbon-13, nitrogen-15, deuterium, and high-field proton nmr, applied to chemical and biological problems. He was Head/Senior Fellow of the

National Nuclear Magnetic Resonance Centre at the ANU.

He is author/co-author of over 100 papers in refereed international journals, along with over 115 articles of more general scientific and technological publications, including interviews with distinguished Professors Carl Djerassi, Chennupati Jagadesh and Paul Anastas.

Finally, Alan practices fine-art, focussing on landscapes in oils, and has won several major prizes for his work. From 2010, Alan was an active contributing member on the Executive and later President (2013 – 2018) of the Artists Society of Canberra.

OTHER BOOKS. CONTRIBUTIONS TO BOOKS

Investing in Knowledge Capital: Management Imperatives, Shantha Liyanage and Alan J. Jones, Hardcover, Singapore Institute of Management, Singapore, January 2002.

Australian Nobel Laureates: Adventures in Innovation, Hardcover, ETN Communications, Sydney, 2004. (Advisor and participant author).

Serendipitous and Strategic Innovation – A Systems Approach to Managing Science-based Innovation, S. Liyanage, J. Annerstedt, P. Gluckman, D. Hunyor, A.J. Jones, and M. Wilson, Hardcover, Praeger Publishers, USA, 2006.

Short Stories; Some of Life's Ups and Downs, Kindle/Amazon Books, USA, November, 2021, Paperback, 123 pages.

13th Corps, A Short History, 1st Battalion North Staffordshire Rifle Volunteers 1858 – 1900, Kindle/Amazon Books, USA, January 2022, Paperback, 131 pages.

Alan is also the author of a series of monographs covering strategic areas of science and their technological applications in industry, including Lasers, Membranes, Biomaterials (for Medical Devices), Superconductivity, Electrodes, Sensors, and Water and Wastes.

Added are comprehensive Reviews of the National Innovation Systems of Finland, Taiwan and Germany; which he visited for these projects. He also provided a one-page *Monthly Column*

on NETWORKS for the journal
Chemistry in Australia, 2009 – 2013.

NOTES

www.ingramcontent.com/pod-product-compliance
Lightning Source LLC
Chambersburg PA
CBHW050535280326
41933CB00011B/1599